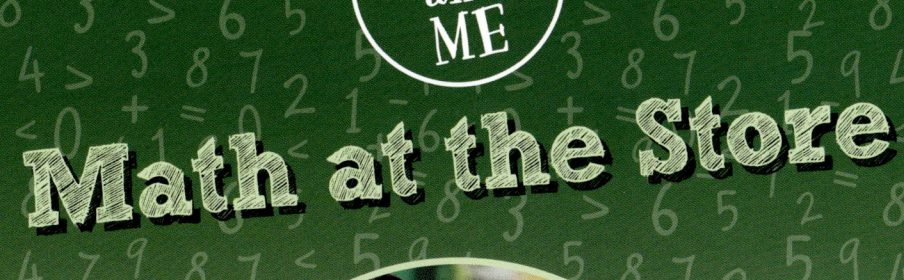

Math at the Store

by Joanne Mattern

Red Chair Press   Egremont, Massachusetts

Look! Books are produced and published by Red Chair Press:

Red Chair Press LLC   PO Box 333   South Egremont, MA 01258-0333

 FREE Educator Guides at www.redchairpress.com/free-resources

Publisher's Cataloging-In-Publication Data
Names: Mattern, Joanne, 1963- author.
Title: Math at the store / by Joanne Mattern.

Description: Egremont, Massachusetts : Red Chair Press, [2022] | Series: LOOK! books : Math and Me | Interest age level: 005-008. | Includes index and suggested resources for further reading. | Summary: "Math isn't just something you learn in school. It's an important part of the world around you. How many objects are in the shopping basket? How many more do we need from our list? How much does each item cost? Young readers will enjoy helping with this shopping trip as they practice counting, adding, and using money"--Provided by publisher.

Identifiers: ISBN 9781643711300 (hardcover) | ISBN 9781643711362 (softcover) | ISBN 9781643711423 (ePDF) | ISBN 9781643711485 (ePub 3 S&L) | ISBN 9781643711546 (ePub 3 TR) | ISBN 9781643711607 (Kindle)

Subjects: LCSH: Mathematics--Juvenile literature. | Shopping--Mathematics--Juvenile literature. | CYAC: Mathematics. | Shopping--Mathematics.

Classification: LCC QA40.5 .M383 2022 (print) | LCC QA40.5 (ebook) | DDC 510--dc23

Library of Congress Control Number: 2021945365

Copyright © 2023 Red Chair Press LLC
RED CHAIR PRESS, the RED CHAIR and associated logos are registered trademarks of Red Chair Press LLC.

All rights reserved. No part of this book may be reproduced, stored in an information or retrieval system, or transmitted in any form by any means, electronic, mechanical including photocopying, recording, or otherwise without the prior written permission from the Publisher. For permissions, contact info@redchairpress.com

Photo credits: Cover, p.1, 3–15, 17–24: iStock; p. 16, 19: Shutterstock

Printed in United States of America
0422 1P CGF22

Table of Contents

Let's Go to the Store 4
Big and Little 14
Finding Your Favorite 16
At the Checkout 18
Words to Know 23
Learn More at the Library 23
Index 24
About the Author 24

Let's Go to the Store

It's fun to go to a store. You can help your mom and dad buy things you need. You use math in many ways as you shop. Let's find out some ways you use math at the grocery store.

A grocery store is a type of store that sells food items like fruits, vegetables, and frozen foods.

The first stop is the **produce** section. You need some apples. You put six apples in a bag. Let's count.

MATH FACT!

Some items in the produce section are priced by weight. So 1 apple might weigh one-half pound. How much would 6 apples cost if the price is $2 per pound?

6 apples = 3 pounds

3 pounds × $2 = $6

Apples are 50 cents each. How much money will you need to buy six apples? Think about ways to group 50 cents six times.

> **MATH FACT!**
>
> 50 cents + 50 cents = $1
>
> $1 + $1 + $1 = $3
>
> Six apples cost $3.

You need to buy bananas. You want enough so you, your mom, and your sister can each have three bananas. How many bananas do you need to buy? Let's count.

|1|2|3| for you.

|4|5|6| for Mom.

|7|8|9| for your sister.

You need nine bananas.

Workers in a store will help you.

This bunch has 12 bananas. How many should you take away to leave nine bananas?

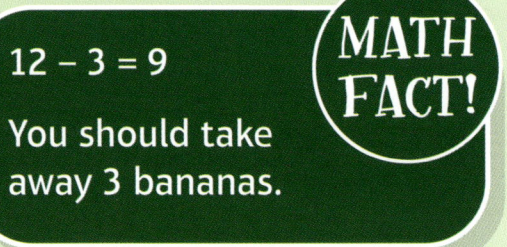

12 minus 3 is 9.

Big and Little

Next you need a bag of carrots.
A small bunch has 6 carrots.
A large bunch has 12 carrots.
Which bunch has more carrots?

MATH FACT!

12 > 6 is how you write 12 is greater than 6. The large bunch has more carrots. Now you know how to write that!

Finding Your Favorite

Here is the cereal aisle. Your favorite is the fourth box from the left. Count the boxes.

1 2 3 4

There it is!

At the Checkout

It's time to pay. Your **groceries** cost $15. You give the **cashier** $20. How much change will you get back?

MATH FACT!

$20 − $15 = $5.

You will get $5 back. You say this as "Twenty minus fifteen equals 5."

Mom says you can have a treat. You choose a pack of cupcakes. There are 6 cupcakes in the pack. Mom says you can have 1 cupcake now. How many cupcakes are left?

6 – 1 = 5

You have 5 cupcakes left.

MATH FACT!

Shopping can be a lot of fun! It takes a lot of math to have a successful shopping trip!

Words to Know

cashier: a person who takes payments in a store

groceries: food and supplies sold in a store

produce: fruits and vegetables

Learn More at the Library

Check out these books to learn more.

Levit, Joe. *Let's Explore Math (Bumba).* Lerner Publications, 2019.

Overdeck, Laura. *Bedtime Math Series.* Feiwel and Friends, 2013.

Reynolds, Mattie. *Super-Smart Shopping (Start Smart: Money).* Red Chair Press, 2013.

Steffora, Tracey. *Sorting at the Market.* Heinemann-Raintree, 2011.

Index

cashier 18

cereal 16

cupcakes. 21

grocery store 4

money. 8

produce. 7

About the Author

Joanne Mattern is the author of many books for children. She loves writing about sports, animals, and interesting people. Mattern lives in New York State with her family.